What's Left?

by April Barth

Today, we went to a street fair. We saw many people and many things. First, we saw four drummers.

Then one drummer walked away, so there were three drummers left.

$$4 - 1 = 3$$

▲ Four minus one is equal to three.

Next, we saw some great dancers.
At first, there were seven dancers.

After a while, two dancers sat down. Five dancers kept dancing.

$$7 - 2 = 5$$

▲ Seven minus two is equal to five.

We started to feel hungry, so we had ice cream. We paid for five ice cream cones.

When one ice cream cone fell, there were four ice cream cones left. We had to buy more.

$$5 - 1 = 4$$

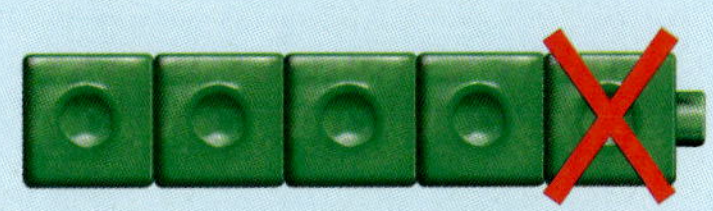

▲ Five minus one is equal to four.

We counted the animals at the street fair. We saw many different kinds of pets.

We saw a man with six dogs! Some were brown and some had spots. Four dogs were brown. How many dogs had spots?

$$6 - 4 = 2$$

▲ Six minus four is equal to two.

People were selling things at the fair, too! First, we saw a woman selling baskets.

We saw eight baskets. Four baskets were large. How many baskets were not large?

$$8 - 4 = 4$$

▲ Eight minus four is equal to four.

Next, we saw a man using wood to make animals. We liked the bears.

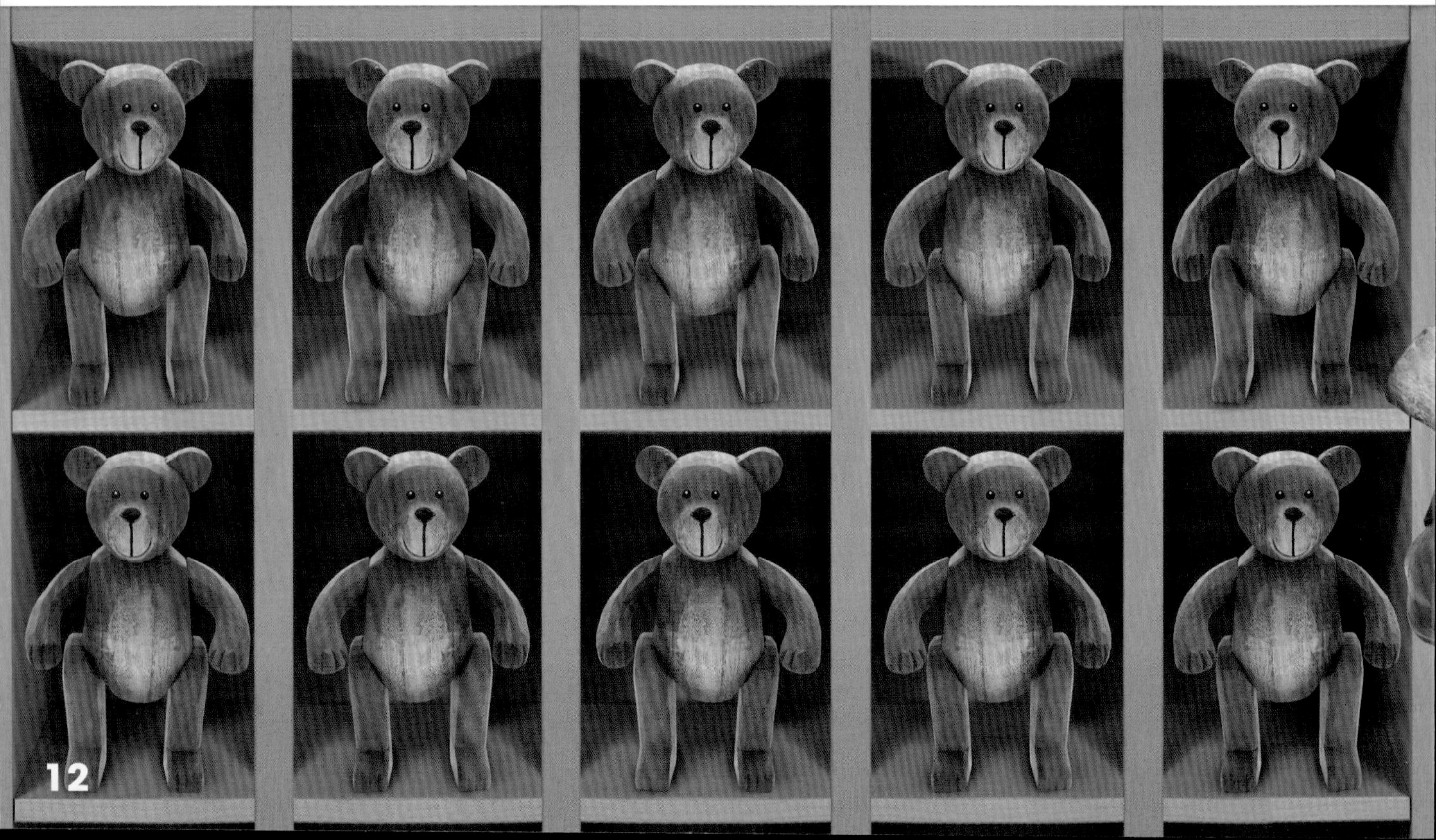

The man was selling ten bears. After we paid for three bears, how many bears were left?

$$10 - 3 = 7$$

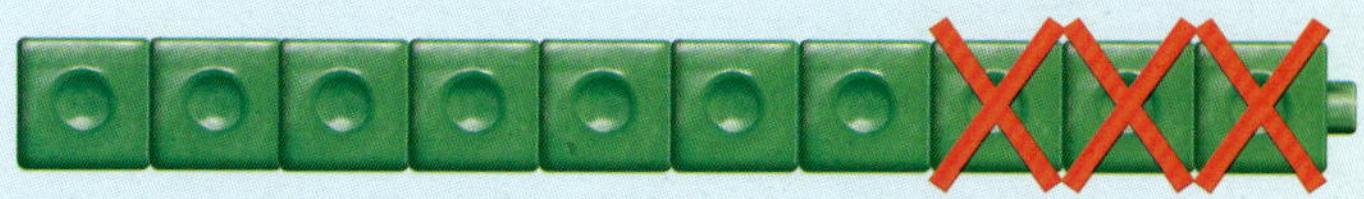

▲ Ten minus three is equal to seven.

We also saw ten necklaces.
The necklaces had beautiful beads.

Five necklaces had blue beads. The rest had red beads. How many necklaces had red beads?

10 – 5 = 5

▲ Ten minus five is equal to five.

We went home after dinner.
We had fun! We want to go
to the street fair again next year.